To Ava

Written, Illustrated and Published by Erica L. Clymer
Royersford, PA USA

For more information visit www.ericalclymer.com

Library of Congress Control Number: 2020902497
ISBN: 978-1-7346063-1-7

Finn's Fun Farm Adventure

Written & Illustrated by Erica L. Clymer

This morning, Finn rises early with the sun.
He's off to the farm for a day filled with fun.

Finn peddles along to the adventure that awaits.
What will he find behind the farm gates?

Finn follows the path through rolling green hills.

At the tippy top, he sees two whirling windmills.

At the barn, Farmer Fred smiles and holds up his rake.
Finn greets him with "Hello, Sir!" and a firm handshake.

"Howdy, Finn!" says Farmer Fred. "Take a look around."
"You'll see at the farm, there is much to be found."

Finn enters the barn filled with bales of hay.

At the stable, the horse greets him with a "Neigh!"

The rooster sings out a "Cock-a-doodle-doo!"

A burly cow belts out a bellowing "Moo!"

The three little pigs roll around in their pen.

Four baby chicks snuggle up to mother hen.

Finn gets quite tired as he counts the sheep.

Up in the trees, bats hang sound asleep.

Buzzing bees swarm around a honeycomb hive.

"Gobble, gobble!" is heard when the turkeys arrive.

Finn plays fetch with Farmer Fred's dog.

Out of the pond leaps a green, speckled frog.

The llama and alpaca look a lot like cousins.

In the coop, clucking chickens hatch eggs by the dozens.

"Woo-hoo!" cheers Finn at the tortoise and hare race.

With a cat on its tail, the mouse picks up the pace.

Finn looks out over the farm's scenic landscape.
"Oh no!" he spots a sheep that made an escape.

Finn dashes up the hill, but the sheep darts away.
In the distance, he hears a trot and a "Neigh!"

The horse gallops over to give Finn a ride.
He hops on and they hurry down the hillside.

Finn and the horse brake to a stop.
They catch the sheep right before a cliff drop.

Heading back to the farm, Finn shows them the way.
In his fun farm adventure, Finn saves the day!

The End

Made in the USA
Coppell, TX
28 May 2020